Fossil Discoveries Disprove Evolution Beyond A Doubt

By

Josh Greenberger

Fossil Discoveries Disprove Evolution Beyond A Doubt

- FOREWORD -

This book shows how the very fossil record that scientists often present as proof of the validity of the theory of evolution proves the precise opposite. Fossils are excellent "laboratory experiments" on a grand scale that assess what happened to life in the past. And they prove beyond any question that the emergence and progression of life did not occur in any way remotely resembling what is depicted by the theory of evolution. This may be a difficult pill for some people to swallow, but it's the facts that should dictate what science is. This book deals with provable facts, not unverifiable conjecture.

The Author

Josh Greenberger's letters and articles -- ranging from science to computers to politics to humor -- have appeared in the New York Daily News, the New York Post, the Jewish Press, Hamodia, and several other publications.

He is the author of "The V-Bang: How The Universe Began," a book that proposed a new big bang theory that seems to solve many cosmological puzzles.

He was the science writer for several years for the publication "Chai Today" and has been a science enthusiast most of his life.

As a computer consultant, expert in over a half dozen programming languages, he has designed and written software for NASA's Goddard Institute of Space Studies, Goldman Sachs, Chase Manhattan Bank, Charles Schwab and several other firms.

www.EvolutionDisproven.com

ISBN-13: 978-1481098052
ISBN-10: 1481098055

v-2

Fossil Discoveries Disprove Evolution Beyond A Doubt

The Fossil Record Disproves Darwinian Evolution

Although animal groupings comprise Phyla, Classes, Orders, Families, Genera and Species, this treatise focuses on life forms with vastly different forms or structures, regardless of their classifications. My use of terms like "different species" and "speciation," therefore, generally refer to life forms that are very different. Life forms that have relatively minor adaptive differences, even if they are technically different species, are not the subject of this treatise.

The scientific concept of the origin of life on earth begins with the premise that life first appeared billions of years ago with the formation of microscopic organisms out of inanimate matter. In the billions of years that followed, small organisms evolved into higher and more complex forms of life through random mutations, and one species evolved into another.

Over the years, a process referred to as "natural selection," scientists believe, weeded out those mutations and organisms less fit to survive than others. Thus, it was mostly the more "fit" that passed on their genetic character traits to subsequent generations. And that's how we and all other life forms got here.

On the surface, this sounds great. However, a deeper analysis of the underlying mechanism and the fossil record, leaves little doubt that a random process, of mutation or any other kind, could not possibly have been the driving force behind the development of life on earth.

First, it should be pointed out that the purported mechanics of speciation are not exactly based on strong empirical evidence, to begin with, as explained on the website of The Department of Geology of The University of California, which has one of the top 25 Geology programs in the country, according to 'America's Best Graduate Schools' by U.S. News and World Report:

"The process of speciation has been difficult to observe, however, and there is still a great deal of controversy about the mechanisms of speciation. No one doubts that it occurs frequently, at least on a geological time-scale. No one has seen a new species form in ecological time, although some cases come very close. You would expect, then, that the geological record, which is so much longer and more incomplete, would hardly ever sample speciation events. We need to include that fact in any theory of speciation. In fact, then, both biologists and paleontologists must infer what happens, and it is very difficult to sort out where fact ends and where interpretation begins. Possibly the term 'speciation' may cover a broad spectrum of events: we already know that some species differ by as few as three genes from others, a difference that would be less than brother-sister differences in other organisms ... Notice that since biologists have not seen a speciation event that everyone would believe, biologists are driven to theory-heavy models of speciation, rather than a rich store of observational evidence. Even so, there are cases of near-speciation in the biological world, and many of them have been ignored because they suggested the 'wrong' answer!"

In addition to showing how the scientific concept of speciation is not exactly based on solid evidence, the above paragraph also shows how dishonest and misleading some scientific literature can get when it comes to evolution.

The University's literature above actually begins with a factual-sounding declaration which I deliberately left out: "The fossil record tells us that new species have evolved from pre-existing ones."

Really?

With all the difficulties presented within the same literature, does the fossil record really tell us that? How can it make a bold statement like, "No one doubts that [speciation] occurs frequently," when the entire paragraph expresses anything but certainty?

The problem with the purported mechanics of Darwinian evolution, though, goes far beyond the lack of evidence for frequent speciation. The lack of an essential by-product of frequent speciation, a long series of happenstance events, completely undermines the fundamentals of Darwinian evolution.

People often challenge the theory of evolution on the basis of whether a random process can produce organization. An analogy often given is: Can an infinite number of monkeys on typewriters, given enough time, produce the works of Shakespeare purely by random keystrokes? Let's assume for the purpose of this discussion that this is possible -- random mutations can, given enough time, eventually produce the most complex forms of life.

Let's get an idea of how that would work by rolling a die (one "dice"). To get a "3," for example, you'd have to roll the die an average of six times (there are six numbers, so to get any one of them would take an average of six rolls). Of course, you could get lucky and roll a 3 the first time. But as you keep rolling the die, you'll find that the 3 will come up on average once every six rolls.

The same holds true for any random process. You'll get a "Royal Flush" (the five highest cards, in the same suit) in a 5-card poker game on average roughly once every 650,000 hands. In other words, for every 650,000 of mostly lesser hands and meaningless arrangements of cards, you'll get only one Royal Flush.

Multi-million dollar lotteries are also based on this concept. If the odds against winning a big jackpot are millions to one, what will usually happen is that for every game where one person wins the big jackpot with the right combination of numbers, millions of people will not win the big jackpot because they picked millions of combinations of meaningless numbers. To my knowledge, there hasn't been a multi-million dollar lottery yet where millions of people won the top prize and only a few won little or nothing. It's always the other way around. And sometimes there isn't even one big winner.

Now, let's take this well-understood concept of randomness and apply it the story of monkeys on typewriters. As mentioned earlier, for the purpose of this discussion we'll assume that if you allow monkeys to randomly hit keys on a typewriter long enough they could eventually turn out the works of Shakespeare. Of course, it would take a very long time, and they'd produce mountains and mountains of pages of meaningless garbage in the process, but eventually (we'll assume) they could turn out the works of Shakespeare.

For simplicity sake, we'll use a limited number of moneys. (My point actually becomes stronger when you use an infinite number of monkeys.)

Let's say, after putting a monkey in front of a typewriter to type out Shakespeare, you decide you also want a copy of the Encyclopedia of Britannica. So you put another monkey in front of another typewriter. Then, you put a third monkey in front of third typewriter, because you also want a copy of "War And Peace." Now you shout, "Monkeys, type," and they all start banging away on their typewriters.

You leave the room and have yourself cryogenically frozen so you can come back in a few million years to see the results. (The monkeys don't have to be frozen. Let's say they're an advanced species; all they need to survive millions of years is fresh ink cartridges.)

You come back in a few million years and are shocked at what you find. What shocks you is not what you see, but what you don't see. First, you do see that the monkeys have produced the works of Shakespeare, the Encyclopedia of Britannica and "War and Peace." But all this you expected.

What shocks you is that you don't see the mountains of papers of meaningless arrangement of letters that each monkey should have produced for each literary work. You do find a few mistyped pages here and there, but they do not nearly account for the millions of pages of "mistakes" you should have found.

And even if the monkeys happened to get all the literary works right the first time, which is a pretty impossible stretch of the imagination, they still should've typed out millions of meaningless pages in those millions of years. (There's no reason for them to stop typing.) Either way, each random work of art should have produced millions upon millions of meaningless typed pages.

This is precisely what the problem is with the Darwinian theory of evolution.

A random process, as depicted by Darwinian evolution and accepted by many scientists, even if one claims it can produce the most complex forms of life, should have produced at least millions of dysfunctional organisms for every functional one. And with more complex organisms (like a "Royal Flush" when compared to a number 3 on a die), an even greater number of dysfunctional "mistakes" should have been produced (as there are so many more possibilities of "mistakes" in a 52-card deck than a 6-sided die).

The fossil record should have been bursting with millions upon millions of completely dysfunctional-looking organisms at various stages of development for the evolution of each life form. And for each higher life form -- human, monkey, chimpanzee, etc. -- there should have been billions of even more "mistakes."

Instead, what the fossil record shows is an overwhelming number of well-formed, functional-looking organisms, with an occasional aberration. Let alone we haven't found the plethora of "gradually improved" or intermediate species (sometimes referred to as "missing links") that we should have, we haven't even found the vast number of "mistakes" known beyond a shadow of a doubt to be produced by every random process.

That randomness will always produce chaos in far greater ratios than anything else, even in cases where it can occasionally produce order of any kind, is an established fact. A process that produces organization without the expected chaos is obviously following a predetermined course.

The notion that the fossil record supports the Darwinian theory of evolution is as ludicrous as saying that a decomposed carcass proves the animal is still alive. It proves the precise opposite. The relative scarcity of deformed-looking creatures in the fossil record proves beyond any doubt that if massive speciation occurred it could not possibly have happened through a random process.

In response as to why we don't see the massive "mistakes" in the fossil record, some scientists point out that the genetic code has a repair mechanism which is able to recognize diseased and dysfunctional genetic code and eliminate it before it has a chance to perpetuate abnormal organisms.

Aside from this response not solving the problem, as I will point out soon, it isn't even entirely true. Although genetic code has the ability to repair or eliminate malfunctioning genes, many diseased genes fall through the cracks anyway. There are a host of genetic diseases -- hemophilia, various cancers, congenital cataract, spontaneous abortions, cystic fibrosis, color-blindness, and muscular dystrophy, just to name a few -- that ravage organisms and get passed on to later generations, unhampered by the genetic repair mechanism. During earth's history of robust speciation through, allegedly, random mutations, far more genes should have fallen through the cracks. Where are they?

And, as an aside, how did the genetic repair mechanism evolve before there was a genetic repair mechanism? And where are all those millions of deformed and diseased organisms that should've been produced before the genetic repair mechanism was fully functional?

But all this is besides the point. A more serious problem is the presumption that natural selection weeded out the vast majority of the "misfits."

A genetic mutation that would have resulted in, let's say, the first cow to be born with two legs instead of four, would not necessarily be recognized as dysfunctional by the genetic repair mechanism. (I'll be using "cow" as an example throughout; but it applies to just about any organism.) From the genetic standpoint, as long as a gene is sound in its own right, there's really no difference between a cow with four legs, two legs, or six legs and an ingrown milk container. It's only after the cow is born that natural selection, on the macro level, eliminates it if it's design is not fit to survive.

It's these types of mutations, organisms unfit to survive on the macro level, yet genetically sound, that should have littered the planet by the billions.

Sure these deformed cows would have gotten wiped out quickly by natural selection, since they had no chance of surviving. But that's precisely the point: Where are all those billions of life forms that were genetically sound but couldn't make it after birth?

How many millions of dysfunctional cows alone, before you even get to the billions of other species in earth's history, should have littered the planet and fossil record before the first stable, functioning cow made its debut? If you extrapolate the random combinations from a simple deck of cards to the far greater complexity of a cow, we're probably talking about billions of "mistakes" that should have cluttered planet earth for just the first functioning cow.

Of the fossils well-preserved enough to study, most appear to be well-designed and functional-looking. Did nature miraculously get billions of species right the first time? With the ratio of aberrant looking fossils being no more significant than common birth deformities, there seems to have been nothing of a random or accidental nature in the development of life.

And to admit that life was not a random process, as I've heard some evolutionists do, and then just leave the question open as to how life got to its current state of diversity, is absolutely absurd and grossly dishonest. There are no other options: it was either an accident or deliberate. And if it obviously wasn't an accident, it had to be by intelligent design.

One absurd response I got from a molecular biologist as to why a plethora of deformed species never existed was: There is no such thing as speciation driven by deleterious mutation.

This is like, upon asking, "How come no one ever leaves the lecture hall through exit 4?" getting a response like, "Because people don't leave the lecture hall through exit 4." Wasn't that the question?

What evolutionists have apparently done is looked into the fossil record and found that new species tend to make their first appearance as well-formed, healthy-looking organisms. So they made a rule out of it: "Speciation is not driven by deleterious mutations." So now that's it's a rule of evolution, you can no longer ask why? If I told you a "rule" that shoes grow on apple trees, can you no longer ask how that works, because it's a rule?

Instead of asking themselves how can a random series of events, which is known to always produce chaos, seldom produce chaos in nature, they've simply formulated a rule in evolutionary biology: There is no such thing as speciation driven by deleterious mutation. This hardly addresses the issue.

It's one thing for the genetic code to spawn relatively flawless cows today. Perhaps, after years of stability, one might argue, nature finally got it right by passing down mostly the beneficial genes. But before cows took root, a cow with three legs, for example, would have been no more genetically deleterious than a cow with four legs. The genetic repair mechanism may recognize "healthy" or "diseased" genetic code, but it can't know how many legs, horns or ears a relatively new species should have, if we're talking about a trial-and-error crapshoot. If the genetic repair mechanism could predict, years before natural selection on the macro level had a chance to weed out the unfit, what a functioning species should eventually look like, we'd be talking about some pretty weird, prophetic science.

In a paper published in the February 21, 2002, issue of Nature, Biologists Matthew Ronshaugen, Nadine McGinnis, and William McGinnis described how they were able to suppress some limb development in fruit flies simply by activating certain genes and, with additional mutations, suppress all limb development during embryonic development.

In another widely publicized experiment, mutations induced by radiation caused fruit flies to grow legs on their heads.

What these experiments showed is how easy it is to make drastic changes to an organism through genetic mutations. Ironically, although the former experiment was touted as supporting evolution, they both actually do the opposite.

The random process that supposedly resulted in such a massive proliferation of life forms on earth could've have created chaos by simply flipping of few genetic "switches." But it didn't even do that! Obviously, the proliferation of life is not the result of random events, neither on the genetic level nor the macro level.

Evolutionists tend to point out that the fossil record represents only a small fraction of biological history, and this is why we don't find all the biological aberrations we should. The issue here, though, is not one of numbers but of proportions.

For every fossil of a well-formed, viable-looking organism, we should have found an abundance of "strange" or deformed ones, regardless of the total number. What we're finding is the proportional opposite.

The theory of evolution may have made sense in the scientifically ignorant days of Darwin. But in the 21st century, evolution appears to be little more than a figment of imagination. Although this imaginative concept has in the years since Darwin amassed a fanatical cult-like following, there is much evidence that contradicts it.

An article entitled, "The Chaos Theory of Evolution," by Keith Bennett, on NewScientist.com, October 18, 2010, describes research that shows the cornerstones of evolution -- adaptation and natural selection -- have little to do with speciation.

Keith Bennett's bio: Professor of late-Quaternary environmental change at Queen's University Belfast, guest professor in palaeobiology at Uppsala University in Sweden, and author of "Evolution and Ecology: The Pace of Life" (Cambridge University Press). He holds a Royal Society Wolfson Research Merit Award.

Excerpts from his article:

"In 1856, geologist Charles Lyell wrote to Charles Darwin with a question about fossils. Puzzled by types of mollusc that abruptly disappeared from the British fossil record, apparently in response to a glaciation, only to reappear 2 million years later completely unchanged, he asked of Darwin: 'Be so good as to explain all this in your next letter.' Darwin never did.

"To this day Lyell's question has never received an adequate answer. I believe that is because there isn't one.

"...the neat concept of adaptation to the environment driven by natural selection, as envisaged by Darwin in 'On the Origin of Species' and now a central feature of the theory of evolution, is too simplistic. Instead, evolution is chaotic.

"Our understanding of global environmental change is vastly more detailed [today] than it was in Lyell and Darwin's time. James Zachos at the University of California, Santa Cruz, and colleagues, have shown that the Earth has been on a long-term cooling trend for the past 65 million years. Superimposed upon this are oscillations in climate every 20,000, 40,000 and 100,000 years caused by wobbles in the Earth's orbit. "

Their research, mostly on birds, "shows that new species appear more or less continuously, regardless of the dramatic climatic oscillations of the Quaternary or the longer term cooling that preceded it.

"The overall picture is that the main response to major environmental changes is individualistic movement and changes in abundance, rather than extinction or speciation. In other words, the connection between environmental change and evolutionary change is weak, which is not what might have been expected from Darwin's hypothesis.

" ... macroevolution may, over the longer-term, be driven largely by internally generated genetic change, not adaptation to a changing environment."

The gist of Bennett's article is that we cannot predict the course of the evolution of life because adaptation and natural selection -- the bedrock of Darwinian evolution -- have little to do with speciation.

But, you may ask, if Bennet's research shows that speciation is driven by some innate genetic characteristics rather than chaotic climate conditions, aren't we back to square one?

No, we're not. Evolution driven by an innate ability of genes to mutate and evolution driven by unpredictable climactic conditions are totally different animals (no pun intended), as will become clear soon.

Genetically driven speciation is analogous to, say, randomly hitting a ball on a billiard table. When the ball drops into a pocket it may have dropped into a random pocket but this was not necessarily a truly random event. The ball can only drop into one of the six pockets available; it cannot drill a new pocket at a random spot.

The point is, the ball can only drop into a pocket that was previously prepared for it, limiting its randomness by a predetermined set number of possibilities. So, no matter how randomly the ball is hit, its "randomness" is limited and guided by the predesign of the billiard table.

This is what I believe is behind speciation. Organisms only change into "allowable," or perhaps genetically guided, life forms. The appearance of a new organism may be a random choice among several "allowable" life forms, but, aside from the occasional aberration, which never results in a lineage of aberrations, an organism will never turn out to be a truly randomly constructed creature.

Fossil records and lab experiments seem to support this type of "organized evolution", which we will name Focused Biological Evolution (FBE), to differentiate it from Darwinian evolution.

Some years ago I read an article about how scientists found a cactus in the desert that had mutated under extreme conditions into another type of cactus. They decided to experiment to see how many different mutations of cacti they could get out of the original one. So they subjected the original cactus to the same conditions that had resulted in it mutating. To their amazement, no matter how many times they performed the experiment, the cactus only changed into that one mutated form.

The scientists in this experiment did not get a myriad of dysfunctional mutations before getting a functioning cactus. They didn't even get several different functioning cacti. The only result was this one mutation, and there seemed to be nothing random about it.

In 2006, a team of researchers from Panama, Colombia and the UK recreated the Heliconius heurippa butterfly in the laboratory by crossing two other species of butterfly, Heliconius cydno and Heliconius melpomene. The process of creating one new species out of two is known as hybrid speciation. Experimenter Chris Jiggins of the University of Edinburgh told BBC News: "The fact [that] we've recreated this species in the lab provides a pretty convincing route by which the natural species came about."

Although this was a "reverse" type of evolution, that the genetic code was able to create a new functional species is an indication of how the genetic code holds some sort of "guidance system" that not only maintains the viability of its host's current form but also that of other forms, and true randomness has little to do with speciation.

In another experiment, in 2002, biologists at the University of California uncovered genetic evidence that explains how large-scale alterations to body plans in animals can be accomplished through what was described as "simple mutations" in a class of regulatory genes, known as Hox, that act as master switches by turning on and off other genes during embryonic development.

Using laboratory fruit flies and a crustacean known as Artemia, or brine shrimp, the scientists showed how modifications in the Hox gene Ubx suppresses 100 percent of the limb development in the thoracic region of fruit flies, and 15 percent in Artemia.

"This kind of gene is one that turns on and off lots of other genes in order to make complex structures," said one graduate student working in William McGinnis' laboratory. "What we've done is to show that this change alters the way it turns on and off other genes. That's due to the change in the way the protein produced by this gene functions."

What this experiment demonstrated is that even in cases where it would have been very easy for nature to create an immense number of bizarre creatures by the simple random setting of genetic switches, nature apparently got these switch settings right the first time in a vast majority of cases, as is evidenced by the mostly functional looking creatures in the fossil record.

As an aside, what's interesting is the simplistic interpretation given by the graduate student about how switches "make complex structures." Switches do not "make complex structures" or cause things to evolve, just as turning on light switches do not cause electricity, light fixtures or wiring to evolve. Switches merely signal a pre-programmed or pre-determined event to occur between existing components. The components themselves may have taken much design and planning.

For an organism's features to simply pop up or disappear with the flick of a switch, there would have to have been a sophisticated underlying mechanism already in place that assigned specific tasks to specific genetic switches. Rather than showing how "simple" it is for new limbs to "evolve," the above experiment shows how sophisticated biological systems really are, and yet how simple it is to change their course of development. Similarly, turning a computer's switch on and surfing the web, for example, is simple enough for a 10-year-old to do, but those simple acts make use of highly sophisticated research, design and development efforts.

Another experiment, this one by evolutionary biologist Richard Lenski of Michigan State University, showed very clearly that speciation is the result of an underlying genetic design and not chaos and randomness.

For twenty years Lenski cultivated 12 populations of bacteria that originated from one single Escherichia coli (E. Coli) bacterium. After more than 44,000 generations, Lenski noticed a similar pattern in all 12 populations; they evolved larger cells, faster growth rates on the glucose they were fed, and lower peak population densities.

Sometimes around the 31,500th generation, one (and only this one) population suddenly acquired the ability to metabolize citrate, a second nutrient in their culture medium that E. coli normally cannot metabolize. The citrate-using mutants then increased in population size and diversity.

Lenski wondered what would happen if he replayed this

experiment; would the same population evolve in the same way, and would any of the other 11 also evolve. So he turned to his freezer, where he had saved samples of each population every 500 generations, and replayed the experiment.

The replays showed that even when he looked at trillions of cells, it was always the same population that re-evolved, and it always evolved only into that same mutation.

This experiment speaks volumes of speciation's non-randomness. Not only was the end result the same every time this experiment was re-played, but the similarity between the intermediate "chaos" of each culture showed that even what gave the appearance of being chaos was actually part of an organized process.

What's mind-boggling is how some evolutionists saw Lenski's experiments as supporting Darwinian evolution, when in fact it did just the opposite. Here's a comment by an evolutionary biologist at the University of Chicago about Lenski's experiment: "The thing I like most is it says you can get these complex traits evolving by a combination of unlikely events. That's just what creationists say can't happen."

Contrary to what this evolutionary biologist claims, nothing in Lanski's experiment evolved in the Darwinian sense. The entire process, after several runs, became as predictable as the "chaos" of an undeveloped fetus turning into a fully formed human being. That's not evolution. Such events are generally referred to as development, formation, maturation, etc., not evolution.

What Lenski's experiments confirmed is that new mutated life forms are not the result of small, random, beneficial, changes, as described by Darwinian evolution, but a genetic predisposition that allows for very specific, predefined forms of life, very much like my earlier billiard analogy. Furthermore, that the genetic code can hold the blueprint for more than one life form is nothing new. We see this quality in some creatures even today:

 * Caterpillars are crawling creatures that go through a stage called pupa, in which they undergo a complete metamorphosis and emerge as flying creatures, butterflies.

 * Tadpoles are aquatic, gill-breathing, legless creatures that

develop lungs, legs, and other organs to roam on dry land.

 * Some salamanders undergo a metamorphosis which also takes them from an aquatic environment to an air-breathing one.

We call these transformations "metamorphoses," as opposed to evolution, because they happen in front of our eyes and it's obvious that their transformations are guided by an innate genetic mechanism, not by an evolutionary process. Had we seen these creatures transform only in the fossil record, and not in front of our eyes, evolutionist would undoubtedly have hailed these transformations as proof of Darwinian evolution.

Darwinian Evolution (DE)
vs.
Focused Biological Evolution (FBE)

You can probably sum up the differences between Darwinian Evolution and Focused Biological Evolution in a nutshell: After a century and a half, we've found more evidence that contradict DE than support it. FBE, on the other hand, is continually being proven in labs, by the fossil record and by archeological discoveries.

After much digging and analysis, we've found that the progression of life as suggested by Darwin is completely absent from fossil and archeological records. Most conspicuous is the absence of the massive number of deformed and diseased life forms that should have littered earth as a result of a long series of random changes.

The vast majority of life forms in fossil or archeological discoveries give the appearance of being well formed and functional organisms. The evidence that DE never happened is spitting in our faces. In fact, the mere proposal by some scientists of a theory like "punctuated equilibrium" (which says that most species experience little change for most of their history, and then, suddenly, new species appear) accentuates the extent to which scientists are at a loss to find empirical support for DE.

In fact, theories like punctuated equilibrium are typical of evolutionists when confronted with contradictions. They simply make a "rule" out of inexplicable findings and, presto, there's no more need to explain. How does life just pop out of nowhere? "Most species experience little change for most of their history, and then, suddenly, new species appear." That really answers that, doesn't it?

One far-fetched, almost comical, explanation given for punctuated equilibrium is that these creatures evolved elsewhere and only their final forms, somehow, mysteriously, appeared in the location where we found sudden appearances of new species.

But the question remains, how come we always find only the fossils where organisms suddenly appeared in their final form and never where they went through the long evolutionary process? Could it be because that long evolutionary process is a myth?

Scientists then start tinkering in the lab with speciation to prove DE. Instead of finding that speciation produces all sorts of random creatures, which is what you'd expect of a random processes, they find that speciation is more of an "action-reaction" process that generally produces some very well-defined, specific, functional organisms. Apparently, speciation seems no more evolutionary than metamorphosis or gestation, albeit requiring different time scales and circumstances.

A theory like punctuated equilibrium actually makes for more comedy than science. Perhaps we should update punctuated equilibrium to the following:

There is overwhelming evidence suggesting that if you incubate three dozen worms in a solution of amino acids and carbon compounds for approximately one and a half million years they will eventually evolve into the Long Island Railroad. The only problem with this theory is that if this were true some species of fish would have a natural tendency to ride the Long Island Railroad. But fish have never actually been observed commuting between Long Island and Manhattan.

A group of enterprising archaeologists, however, found the missing link to this apparent puzzle. Digging through the ruins of an old Long Island Railroad yard, they came across a fossil of a fish believed to be extinct for billions of years. In fact, after taking a radiocarbon reading of the fossil and the brown paper bag it was found in, they confirmed that their find dated back to the "big bang," give or take six months. This proves conclusively that prehistoric fish did commute via the Long Island Railroad.

Now, the question arises, did prehistoric fish commute on dry land or did prehistoric trains run underwater? No one really knows for sure. But, the famous Dr. Imust Beagenius (pronounced I-must Be-a-genius) is grappling with a theory. Dr. Beagenius suggests that prehistoric fish must have travelled on dry land. He points out that extensive laboratory tests show that railroad tickets are not waterproof.

There you have it -- a theory which links fish, worms, and the Long Island Railroad. It couldn't be more logical.

Unfortunately, not everyone is that easy to please. There are those who, believe it or not, would demand a more detailed explanation of such a theory, no matter how logical it sounds. "How do a bunch of worms," they would naively ask, "turn into the Long Island Railroad?"

In spite of the absurdity of such skepticism, I offer the following evidence which should render this theory proven beyond a shadow of a doubt.

Our archeologist friends went back to the same railroad yard and made some more astonishing discoveries. They lined up some of the old cars side by side and noticed how each car was slightly bigger and better developed than the one before it. The car at one end had a highly sophisticated and powerful air conditioning system, while the car at the other end had not even a fan. The only trace of air conditioning found in one underdeveloped car was the fossil of a conductor slapping an old woman with his cap to create some air disturbance. (His cap, incidentally, has been known to be extinct for at least seven and a half billion years. It had no union label.)

Then, scientists took a worm crawling in the same railroad yard and put it under a powerful electron microscope. And behold, they made an astounding discovery: A worm's cell magnified three billion times has an uncanny resemblance to a train window (without the shades).

It's quite obvious that the evidence presented for the worm-train theory overshadows the somewhat popular but fanatical notion that trains may have been manufactured by intelligent beings. The "intelligent beings" theory would imply a labor union. So far, none of the trains studied showed any traces of major medical benefits, pension funds, or sick leave. How such a ridiculous theory even got started is hard to imagine. So much for this nonsensical "intelligent beings" theory.

By now you must be saying to yourself, "Well, the evidence for the worm-train theory is certainly overwhelming. Any idiot can see its scientific validity. But where did the first worm come from?"

I'm glad you asked. The theory widely accepted by the scientific community and also strongly supported by our famous Dr. Imust Beagenius is the "big bait" theory. In the beginning there was a big ball of fishing hooks. Nature found it rather absurd to have so many fishing hooks without worms. In a few short billions of years, worms began to materialize around the hooks. When the first trout started biting, nature found it necessary to produce more worms to keep up with the fishing season. And so, worms began materializing on virtually every hook around the globe. Then, in the off-season, there were more worms than hooks. So, the problem at that point was storing these excess worms. This brought about the invention of the can. So, you see, the worm-train evolution began with the Big Bait. And the Big Bait began with a can of worms.

How's that for a new theory?

I heard one evolutionist even admit that life could not have been an accident. But he wouldn't acknowledge it must have been intelligently designed. This is quite an absurd position. It's got to be one or the other. Something is either an accident or deliberate; there is no in-between and no other options. And if you prove one, you've disproven the other. Conversely, if you disprove one, you've proven the other.

If all evidence shows clearly that the development of life on earth was not the result of accidental occurrences, that demonstrates conclusively that it had to be intentionally designed. To understand the former but not acknowledging the latter is intellectual dishonesty, at best, delusional, at worst.

How is FBE different?

While Darwinian evolution began as a theory in search of evidence, FBE is a direct result of that evidence. Unlike DE, FBE is not a theory waiting to be proven; it's the evidence that created it. What's more, FBE not only explains the fossil record and speciation in the field and the lab, but, interestingly, it is also fully compatible with Creation.

Here's a capsulized review of how FBE would explain the development of life on earth from its inception to today.

Please note that FBE does not explain how life began. And neither does any other science. There is not a scintilla of empirical evidence in the lab or in the field that shows abiogenesis (living organisms arising from inanimate matter) ever occurred or is even possible. Yet, we are here; something or someone had to have started life. So with the complete absence of any science to explain the beginning of life, using Creation as a model is as good as any.

In the beginning, all of today's ancestral life forms were Created. (Whether "Created" means ex nihilo or that the land and sea gave forth their respective creatures is irrelevant to this discussion.)

As these ancestral life forms spread or appeared throughout various climates around the globe, they went through changes to adapt to their environments and, in some cases, speciation may have occurred.

Being that every known (and perhaps as yet unknown) variation of life has its roots in genetic code rather than accidental occurrences, adaptation and speciation did not require massive trial-and-errors or long development periods. Instead, they were as smooth and as precise transformations as the metamorphosis of tadpoles into frogs and caterpillars into butterflies.

(Speciation involving intermediate chaotic-looking organisms, by the way, has thus far been found only in micro organisms. And even then, the "chaos" always have similarities, with the end result always appearing as a specific genetically-dictated mutation, not as a randomly generated organism.)

The sudden appearance of new species in the fossil record, therefore, is precisely how it must've happened. New species could easily have popped up within a generation or two. For without the need of Darwin's lengthy development period, millions of years of myriads of "misfits" and missing links were not necessary (even if they could possibly evolve life).

As far as scientific explanations go, DE has been a 150-year failure. It's time we discarded DE, as we've done with many other outdated "earth is flat" type of theories. The sophistication of the 21st century calls for a new theory that fits the facts, not an old patched-up theory that has its roots in ignorance and needs a new patch for every discovery. Focused Biological Evolution could be that new theory.

What Qualified Charles Darwin To Propose the Theory of Evolution?

Well, let's look at his background. At the age of 13, Charles Darwin was sent to school to study letters. He failed miserably. At the age of 16, his father used his influence to get Charles accepted into medical school.

But Charles was not cut out for this. In January 1826 Charles had written home complaining of "a long stupid lecture" about medicine. He loathed medicine and left in April 1827 without a degree.

Finally, at the age of 22 Charles Darwin studied and received a degree in Theology.

A degree in Theology?

FOSSIL DISCOVERIES DISPROVE EVOLUTION BEYOND A DOUBT

A degree in Theology qualified Charles Darwin to postulate the theory of evolution? What exactly was his theory based on?

Apparently, Charles Darwin based his theory of evolution on little more than personal observations and subjective reasoning. That is, an entire branch of "science" today is based on the imagination of one person who had no scientific credentials. The average high-school student today knows more about genetics than Charles Darwin knew about it then.

What's even stranger is that a contemporary of Darwin, Gregor Mendel, was more qualified than Darwin to speak of biological life and challenged Darwin's views. Yet, Mendel's views never took hold in a big way, and much of his work was not even recognized until after his death.

Darwin assumed that there were no limits to biological variation and that, given enough time, a fish could eventually evolve into a human being. Gregor Mendel challenged this assumption, claiming evolution was restricted to within the "kinds." That is, Mendel maintained that a life form could evolve into something related to its own "kind," but a drastic development such as a fish evolving into a human being, no matter how much time was allowed, could never happen.

Was Mendel's version of evolution not accepted because he was less qualified to speak about biological life than someone holding a degree in theology?

Well, what was Mendel's background? Mendel was an Austrian biologist whose work on heredity became the basis for modern genetics. He had a science education at the University of Vienna, and wrote about geology and organic evolution on his 1850 teaching examination.

Unlike Darwin, Mendel's theories were based on extensive research and experimentation, which began in 1856, three years before Darwin published his Origin of Species. Mendel carefully designed and meticulously executed experiments involving nearly 30,000 pea plants followed over eight generations.

In 1866, Mendel published his work on heredity in the Journal of the Brno Natural History Society. However, the importance of his work only gained wide understanding in the 1890s, after his death, when other scientists working on similar problems re-discovered his research. William Bateson, a proponent of Mendel's work, coined the word genetics in 1905.

With all of Mendel's qualifications and achievements, you'd think his version of evolution would have been the one to catch on. After all, archeological discoveries to this day show that Darwin's long progression of slow, incremental, evolutionary changes never happened; archeology could certainly not have supported evolution in those days. But, somehow, it was Darwin who received widespread recognition, not Mendel.

How did this happen?

Apparently, Darwin's theories had more political attraction than scientific substance. Here's an excerpt from the National Institutes of Health, nih.gov, from an article entitled "Theories of evolution shaping Victorian anthropology. The science-politics of the X-Club, 1860-1872:"

It refers to a paper that " ... discusses the role that a group of evolutionists, the X-Club, played in the epistemic and institutional transformation of Victorian anthropology in the 1860s. It analyses how anthropology has been brought into line with the theory of evolution, which gained currency at the same time. The X-Club was a highly influential pressure group in the Victorian scientific community. It campaigned for the theory of evolution in several fields of the natural sciences and had a considerable influence on the modernization of the sciences ... evolutionary anthropology emerged in the 1860s also as the result of science-politicking rather than just from the transmission of evolutionary concepts through discourse."

And, to this day, some of the strongest voices behind evolution argue not from a scientific perspective, but from personal conviction. If you look at evolution blogs you'll find that Darwinian evolution quite often (although not always) goes hand in hand with atheism. Evolution is regularly used by atheist as an intellectual tool for arguing that life took no intelligence to design.

Why attempt to use science to detract from life's obvious inherent design? Well, it's difficult to deny, especially in this day and age, that there is complexity and sophistication in nature. So to deny that life required an intelligent creator, no matter how desperately you'd like to, for whatever personal reasons, just seems illogical and downright idiotic.

But, what if you can come up with a "modern" idea that denies it for you, and, at the same time, makes you look like a "progressive?" Now that's something some people can sink their teeth into. Darwinian evolution is just that vehicle. Is it science? Absolutely not. But in the hands of an atheist, it's an armored tank. One well-known British evolutionary biologist is known more for his rants and lectures against the concept of God than for his discussions on science.

In the final analysis, all evidence points to order and harmony governing every aspect of the development of life. Random external forces may play a role in a new life form emerging, they may also play a role in bringing out certain features that will help an organism survive, but they do not design physical features or the genetic switches that control these features. New features are nothing more than expressions of dormant genetic traits.

Thus, not only is there nothing accidental about the development of life, but the genetic structure, as complex as we've already known it was, appears to be even more complex than anything we've imagined. For the genetic code to hold the key to an organism's current form and also to the forms of several new variations or species is truly mind-boggling. How serious does one's evolution delusion have to be to not see the design in all this?

What's interesting is that DE has more holes in it than the big bang. Yet, you'll occasionally hear scientists admit there are problems with the big bang and question whether it's the correct theory about the beginning of the universe. I even saw one scientist write that he believed in the big bang because "we have nothing better."

Not so with evolutionists. Just about every evolutionists I've encountered is absolutely convinced that DE, despite all evidence against it, is a solid, one-hundred-percent-correct theory. With all the obvious problems with DE, how can one be that sure? The answer is, DE has turned into a cult.

DE evolutionists, I believe, fall into two broad categories. Those who perpetuate the theory and know it has no legs to stand on, and those who don't know better and just rely on "the scientist."

One guy I spoke to recently had exactly that response. He admitted he knew little about science but said he believed in DE because he relied on scientists. Scientist, he reasoned, gave us things like cell phones, heart transplants, Ipads, etc., they must know what they're talking about.

The truth, however, is that the scientists who gave us all of life's conveniences are not necessarily the same ones who perpetuate DE.

Scientists are human (even those who sound like they evolved from apes). Just like there are good doctors and quacks, good lawyers and shysters, good car mechanics and crooks, there are "good" scientists and "junk" scientists. DE evolutionists are the shysters of molecular biology.

I had one debate with someone on an atheist forum who was absolutely convinced about the veracity of DE and claimed he even had a paper by a molecular biologist that proved the correctness of DE. When I examined his paper, and saw that it made little sense, I asked him to explain what he understood about the paper. He couldn't explain any of it.

The paper I believe was written by a molecular biologist, and perhaps it somehow made some sense to him, or perhaps it was deliberately written to confuse, but it was presented as "proof" by someone who had no idea what it said. This approach, I believe, represents the majority of laymen who believe in evolution; they have little knowledge of science but simply take "scientists" word for it.

I later debated the molecular biologist who supposedly wrote this paper. His reasoning went in circles, he clarified nothing, but he had everyone on the forum convinced he was a "superstar" and knew why evolution worked.

The perpetuation of DE also has elements of intimidation. There's a documentary out by a famous actor/comedian who interviews scientists who have been harassed and even fired from universities for suggesting that life could not possibly have evolved without intelligence. Is this what they call a scientific debate? As I've mentioned before, DE is not at all about science. It's a cult with an agenda.

DE also gets much unwarranted traction from the media, which also relies on "the scientists." Here's an article that ran in the New York Times on May 18, 2009:

"On Tuesday morning, researchers will unveil a 47-million-year-old fossil ['Ida'] they say could revolutionize the understanding of human evolution at a ceremony at the American Museum of Natural History.

"But the event, which will coincide with the publishing of a peer-reviewed article about the find, is the first stop in a coordinated, branded media event, orchestrated by the scientists and the History Channel, including a film detailing the secretive two-year study of the fossil, a book release, an exclusive arrangement with ABC News and an elaborate Web site.

"The specimen, designated Darwinius masillae, is of a monkeylike creature that is remarkably intact: even the contents of its stomach are preserved. The fossil was bought two years ago in Germany by the University of Oslo, and a team of scientists began work on their research. Some of the top paleontologists in the world were involved in the project, and it impressed the chief scientist at the Natural History museum enough to allow the press-conference.

"'We would not go forward with this, even in a hosting capacity, unless we had a sense of the scientific importance,' said Michael J. Novacek, the provost of science at the museum.

"'It's the most newsworthy and noteworthy special we've been a part of,' said Nancy Dubuc, the general manager of the History Channel. 'We made a commitment early on to get behind it in a big way: to see it through peer review, and see that it is the media event it should be.'"

This was my response, which was published in the New York Post on May 26, 2009:

"The fossil Ida is being used by scientists as an assault on a gullible public.

"One fossil does not represent a transitional species, any more than the remains of a two-headed snake represents a transition of snakes from one head to two heads. They're simply aberrations of nature.

"You'd need more than one fossil to represent a species, and you'd need many transitional aberrations that couldn't survive to show an evolutionary process was going on.

FOSSIL DISCOVERIES DISPROVE EVOLUTION BEYOND A DOUBT

"Ida represents the fanciful speculations of a scientific community determined to publicize its biased agenda."

On October 22, 2009, the New York Post ran the following article detailing how scientists realized in the end that Ida was just one big mistake:

"Remember Ida, the fossil discovery announced last May with its own book and TV documentary?

"A publicity blitz called it 'the link' that would reveal the earliest evolutionary roots of monkeys, apes and humans. Experts protested that Ida wasn't even a close relative. And now a new analysis supports their reaction.

"In fact, Ida is as far removed from the monkey-ape-human ancestry as a primate could be, says an expert at Stony Brook University on Long Island.

"Professor Erik Seiffert and his colleagues compared 360 specific anatomical features of 117 living and extinct primate species to draw up a family tree. They report the results in today's issue of the journal Nature.

"Ida is a skeleton of a 47-million-year-old cat-sized creature found in Germany. It starred in a book, 'The Link: Uncovering Our Earliest Ancestor,' and a TV documentary narrated by David Attenborough.

"Ida represents a previously unknown primate species called Darwinius. The scientists who formally announced the finding said they weren't claiming Darwinius was a direct ancestor of monkeys, apes and humans. But they did argue that it belongs in the same major evolutionary grouping, and that it showed what an actual ancestor of that era might have looked like.

"The new analysis says Darwinius does not belong in the same primate category as monkeys, apes and humans. Instead, the analysis concluded, it falls into the other major grouping, which includes lemurs.

"The primate skeleton 'Ida,' once called 'the link' to an evolutionary ancestor of humans and apes, turns out not to be even close."

So, that Ida was a link in the evolutionary chain was trumpeted with a ceremony at the American Museum of Natural History, peer-review articles, the History Channel, a film, ABC News, an elaborate Website, some of the top paleontologists in the world and the chief scientist at the Museum of Natural History. In the end, it turned out to be not even close.

What happened in the case of Ida is similar to what happens with many evolutionary claims. The initial claim gets widespread publicity, while the refutations barely make the news.

Ida's demise as an evolutionary link ran in a few articles here and there, but got nowhere near the publicity that Ida's unveiling got. How many people do you think still believe the original hype about Ida? Probably anyone who read or heard the hype but never got wind of the retractions. That's a heck of a lot people. This is how such an empty theory can have such a wide following.

And how did so many "experts" get fooled by a fossil that had no relevance to their claim? Were they all really fooled? They can't all be that incompetent. I don't think they are. Some of them are downright dishonest.

Here's one response I saw on an online forum to my statement that one fossil does not represent a transitional species: " ... scientists have many transitional fossils ... "

Right. Is that why they made such a big deal out of Ida? Do they normally hail the five-thousandth "discovery" of the same thing? Do we have a record of who "discovered" Florida for the five thousandth time? Do we know who "invented" the engine even for the five hundredth time?

Ida received such accolades because scientists knew they had nothing like what they believed Ida represented. If scientists believed they already had evidence of Darwinian evolution, what was the big deal about Ida?

Ida was a big deal because there was no empirical evidence to support Darwinian evolution as late as 2009. And now that Ida has been debunked, DE remains a figment of the imagination, based on no science whatsoever.

(Needless to say, the guy on the forum never presented even one of the many fossils he claimed proved Darwinian evolution.)

In the final analysis, it's not the job of scientists to tell us what science is. It's their job to investigate nature and present their findings. And it is these -- provable -- findings that constitute science.

For scientists to ignore the obvious because it may lead to what in their view is unscientific, is grossly disingenuous and simply not their call. To ignore the obvious fact that life was not the result of accidental events -- a fact supported by almost every fossil ever found -- because the concept of God is not scientific, is really jumping the gun. Scientists do not have to talk about God, if they prefer not too. But they do have an obligation to put forth their honest findings, and let the public decide whether they want to talk about God.

That life shows no signs of being an accident is a simple conclusion and, at that level, does not constitute religion. Not reporting such an obvious conclusion, however, is nothing short of bias and deception.

The sad part is that in this day and age Darwinian evolution is still being taught in school as science. Unfortunately, most of our legislators and school board members are, when it comes to science, laymen. So when evolutionists, some of whom may have accredited degrees, argue in favor of teaching DE in school, how can legislators and school board members argue against it? They really don't have much of a choice.

I'm convinced that if the argument presented in the last chapter, that the fossil record shows absolutely no signs of an accidental evolutionary process, is presented to legislators and educators, and evolutionists are challenged to produce fossils that show otherwise, this cult called Darwinian evolution can be eliminated from the classroom.

Abiogenesis:
Is It Even possible?

In the beginning the Earth was almost formed but void of life, and a primordial soup comprised of water, hydrocarbons and ammonia was upon the face of the deep; and the spirit of abiogeneses hovered over the face of the waters. And a lightning bolt struck the soup, and, behold, the building blocks of life were created.

And there was soup in the evening and lightning in the morning, and this was one theory. And scientists saw that this theory was good and called it science.

Is it me, or does this sound like Creation? The only thing missing is God.

The "scientific" theory of lightning creating the first Amino-acids is as close as science has ever gotten to explaining the initial appearance of the building blocks of life on Earth. How inanimate matter than came to life (abiogenesis), nobody knows.

Nobody knows because no one has ever reproduced abiogenesis and there is no evidence of it ever occurring. So if no one's ever reproduced it and there's no evidence of it occurring, what makes it science? And what makes it better than Creation? That is, if you say that God caused inanimate matter to come to life, that's not science because you can't prove it. But if you say that inanimate matter came to life through some other unprovable process, a process that some scientists even believe may never be possible to prove, that is science. Why?

From the euphoria displayed by scientists every time there is the slightest hint that evidence of abiogenesis is about to be uncovered, and the disappointments that invariably follow, it seems scientists' faith in abiogenesis is based more on emotional expectations rather than meaningful facts.

In April 2007 a team of European astronomers announced that, using a telescope in La Silla in the Chilean Andes, they discovered an Earth-like planet (named Gliese 581c) 20.5 light years away that could be covered in oceans and may support life.

An article on DailyMail.co.uk. reporting on this discovery, using a tactic typical of science writing, begins with, "[Gliese 581c has] got the same climate as Earth, plus water and gravity. [This] newly discovered planet is the most stunning evidence that life -- just like us -- might be out there." The article then admits, "We don't yet know much about this planet," but goes on to say, "This remarkable discovery appears to confirm the suspicions of most astronomers that the universe is swarming with Earth-like worlds."

Stunning evidence that life just like us might be out there? The universe is swarming with Earth-like worlds? Does this discovery really say all this?

Only a month later, dismay set in over Gliese 581c having been erroneously touted as an Earth-like planet. As one website put it: "...the source of so much press speculation about terrestrial worlds, turns out to be far too hot to support life ... it's closer to its star than Venus is to ours." And that was the "end of life" on this "Earth-like" planet.

The practice of publicizing discoveries along with wishful interpretations before facts are checked is common in scientific circles. Then, when facts that contradict initial assumptions come out, they are often not given the same urgency and publicity as the original announcements. The public is thus left with perceptions that coincide with what scientists would like to believe rather than with the way things really are.

Another planet discovered quite close to us in space was described by NASA in April 2004 as follows: "The similarities [to Earth] are striking. Each planet has roughly the same amount of land surface area. Atmospheric chemistry is relatively similar, at least as Earth is compared to ... other planets in [our] solar system. Both planets have large, sustained polar caps and the current thinking is that they're both largely made of water ice. The ... planets also show a similar tilt in their rotational axises, affording each of them strong seasonal variability. [They] also present strong historic evidence of changes in climate."

This planet is Mars.

If we had found a planet so similar to Earth several billion light-years away, scientists would have been screaming with euphoria that we've just about found life on another planet. In fact, at one point we did entertain the thought that Mars may contain life, and the word Martians became a staple of science fiction for many years.

Then what happened? We explored Mars. Suddenly, the Martians disappeared, and we're now down to dredging up soil to find microorganisms. The disappointments in exploring Mars go far beyond bruised egos; they've shaken the very foundation of abiogenesis.

In December of 2007, scientists at the Carnegie Institution's Geophysical Laboratory had shown, by analyzing organic material and minerals in the Martian meteorite Allan Hills 84001, that building blocks of life (organic compounds containing carbon and hydrogen) did form on Mars early in its history.

The Phoenix lander's May 31st, 2008, transmission of photos of ice on Mars was hailed as a possible breakthrough in our search for life on other planets. By July, the Phoenix lander had detected water in the Martian soil. "We have water," proclaimed William Boynton of the University of Arizona, lead scientist for the Thermal and

Evolved-Gas Analyzer (TEGA). "We've seen evidence for this water ice before in observations by the Mars Odyssey orbiter and in disappearing chunks observed by Phoenix last month, but this is the first time martian water has been touched and tasted."

So, after finding the building blocks of life and water, have we found life on Mars? No, we haven't. Why not? The answers you get usually go along the lines of, "We have to dig some more," or, "We've only explored a small portion of Mars."

If you were an alien visiting Earth's vicinity, how many orbits around Earth would you have to make before discovering life? Not even an entire orbit. Half way around Earth you'd discover a plethora of life. Would you even have to land? Of course not; any half decent telescope in orbit would detect life on Earth. And you certainly wouldn't have to dig.

We do know one thing about Mars for just about certain; there is no life on the surface. This alone is a serious problem, as far as biogenesis is concerned. Earth and Mars, according to scientists, were formed in roughly the same period of time and from the same stuff in space, 4.5 billion years ago. During that time Earth has produced literally billions and billions of life forms, some as huge as dinosaurs, some as advanced as humans. Mars, however, in a staggering 4.5 billion years, has produced absolutely no life that we can discern -- not even small ants! How's this possible?

Even if life on Mars had somehow gotten wiped out, we'd at least have to find some bones, carcasses or something. But nothing? What we've found is a planet that seems to be totally barren.

The mere fact that we have to dig in hopes of finding any traces of life on a planet with such strong similarities to and the same age as Earth says there's something wrong with the concept of biogenesis. Ironically, scientists see the discovery of the building blocks of life and water on Mars as hopeful signs of someday finding life there, when in fact the opposite is true. Being that these vital components of life do exist shows very clearly that inanimate matter does not come to life.

And the notion that the Martian environment is too harsh to support life rings pretty hollow. Harsh environments do not deter life here on Earth. Here's an idea of how harsh things can get here on Earth, and how life thrives in spite of it:

In 1977 we found the first hydrothermal vent, an opening where water heated by Earth's molten interior is released into the ocean. Closest to the vent, in the midst of water which sometimes exceeds 450 degrees Fahrenheit, were eight-foot long tube worms. Most animals need sunlight to survive; the area where these tube worms thrive receive no sunlight whatsoever.

Then, as if to laugh in the face of what's considered "normal" for biological life forms, these tube worms had no eyes, mouth, or intestinal tract. They get their nourishment from surrounding bacteria.

To add to this ecological mystery, these bacteria thrived on hydrogen sulphide, which is found in the water coming from the hot vent. To most higher animals, hydrogen sulphide is as poisonous as cyanide!

Since 1977 many more vents have been discovered on the ocean floors. Besides tube worms, other exotic animals have been found thriving in the immediate vicinity of the vents -- pink fish, snails, shrimp, sulphur-yellow mussels, and foot-long clams, to name a few. Similar animal populations have since been discovered in waters only a few degrees cooler than freezing. Talk about adapting to extreme and adverse conditions.

Cacti are known to survive the most difficult and unusual climates. Their ability to sustain themselves in areas of little rainfall, hot dry winds, low humidity, strong sunlight, and extreme fluctuations in temperature is nothing short of phenomenal. Some cacti can survive internal temperatures of near 145 degrees Fahrenheit. Most plants haven't got a chance where some cacti prosper.

Lichens, a combination of fungus and algae, have been found thriving in an area of Antarctica where temperatures sometimes get colder than 70 degrees below zero Fahrenheit. As far as hostile environments go, this seems to be the extreme opposite of deep, dark, hot waters.

Bacteria have been found growing an amazing 25 feet underground in Antarctica.

In the course of Earth's history, there have probably been over a half billion animal species in existence, from such monstrosities as whales and dinosaurs right down to microscopic life forms such as amoebas and viruses. That's a half billion before you even bring plant life into the picture.

The planets in our solar system, according to scientists, formed about four and a half billion years ago. The most primitive forms of life allegedly appeared on Earth as far back as three billion years ago. Huge creatures such as dinosaurs roamed our planet an alleged 200 million years ago, and ruled for an enormously long period of over 100 million years. Finally, scientists believe, humans appeared about two to three million years ago.

That is, something as complex as the human brain has allegedly been around for at least a staggering two million years. An optical instrument as sophisticated as the eye has been around even longer.

Yet, when we look at a planet formed at the same time and from the same stuff as Earth, right next to us in space, what do we find? We find a barren world with absolutely no traces of life. We have to dig in search of even the simplest organism, which we have not yet found. Is there something wrong with this picture?

Sure the Martian environment is hostile. But two miles down at the bottom of our oceans near vents which spew hot water mixed with hydrogen sulphide in total darkness is not exactly a summer vacation spot -- it's about as hostile as an environment can get! But life thrives there in complete defiance of what are normally considered ecological adversities.

So is 25 feet deep in the ice of Antarctica a hostile environment. So is the desert. Furthermore, in that alleged period of three and a half billion years ago, the entire Earth, according to scientists, was hostile. Life on Earth allegedly began in an environment which would be hostile to many of today's life forms. And many of today's life forms live in conditions which would have been intolerable to the organisms which allegedly brought life into existence billions of years ago. But life on Earth thrives in spite of it all.

It's hard to imagine life on Earth being completely wiped out by any natural or manmade disaster. But somehow, life on Mars has either been completely wiped out (and the telltale traces mysteriously hidden) or life on Mars never came into existence. It's totally inconceivable that something as tenacious and as diversified as life has not left its mark on Mars.

Well, maybe there's no life on Mars because the notion of inanimate matter coming to life is a fantasy. It doesn't happen and it's never been proven to happen. Mars actually proves that given billions of years an entire planet will never produce even one single microscopic organism.

It follows logically that if abiogenesis does not work, we may very well be the only life, as we know it, in the universe, which I believe is the case. Again, it is scientists' job to give us honest conclusions based on facts, not interpretations based on biases.

I understand it must be a frightening thought to some scientists, if we're not just some "accident" or "probability" in a universe bursting with billions of civilizations, we may be here by design. But that's for the public to deal with, not for scientists to rule out.

Outdated Dating Methods

What are the methods used by scientists to date archeological finds? And do those methods tell the true age of buried organisms?

The method used by scientists to determine the age of archaeological finds is called radiometric dating. It involves measuring decayed radioactive elements and, by extrapolating backward in time, determining the age of an organism.

One element commonly used, in what's referred to as "radiocarbon dating" or "radiocarbon reading," is C-14, a radioactive isotope of carbon, which is formed in the atmosphere by cosmic rays. All living organisms absorb an equilibrium concentration of this radioactive carbon. When organisms die, C-14 decays and is not replaced. Since we know the concentration of radioactive carbon in the atmosphere, and we also know that it takes 5,730 years for half of C-14 to decay (called a "half-life cycle"), and another 5,730 years for half of what's left to decay, and so on, by measuring the remaining concentration of radiocarbon we can tell how long ago an organism died.

Since C-14 can only give dates in the thousands of years, elements with longer half-life cycles (such as Samarium-147, Rubidium-87, Rhenium-187, Lutetium-176, to name a few, with half-life cycles in the billions of years) are used to date what are believed to be older archaeological finds. The procedure is roughly the same; the amount of decay is measured against the initial amount of radioactive material, giving the object's supposed age.

One obvious flaw in this technique is that we don't really know the level of radioactive concentration acquired by an organism which lived before such recorded history. Scientists make a bold assumption that the atmospheric concentration of the radioactive material -- carbon or any other element -- being measured has not changed since the organism's death.

Another bold assumption made by scientists is that the rate of radioactive decay has remained constant throughout history.

Are these valid assumptions?

Hardly.

In 1994 Otto Reifenschweiler, a scientists at the Philips Research Laboratories in The Netherlands, showed that the radioactivity of tritium could be reduced by 40 per cent at temperatures between 115 and 275 Celsius. That is, under certain conditions, the environment can effect radioactive decay.

In 2006 Professor Claus Rolfs, leader of a group of scientists at Ruhr University in Bochum, Germany, in an effort to reduce nuclear waste radioactivity, has come up a with a technique to greatly speed up radioactive decay. Rolfs: "We are currently investigating radium-226, a hazardous component of spent nuclear fuel with a half-life of 1600 years. I calculate that using this technique could reduce the half-life to 100 years. At best, I have calculated that it could be reduced to as little as two years ... We are working on testing the hypothesis with a number of radioactive nuclei at the moment and early results are promising ... I don't think there will be any insurmountable technical barriers."

Reducing 1600 years to two years is a phenomenal 98 percent reduction. This means that an archeological find that has gone through environmental conditions similar to those in the lab could appear to be 300,000 years old when in fact it's only six thousand years old.

What's more, if scientists, with relatively limited resources, can speed up radioactive decay 800 times, the violent upheavals of earth's history could certainly have sped up radioactive decay by far greater numbers. Thus, if radioactive decay increased, say, 1 million fold, an organism thought to be 4 billion years old, based on today's rate of radioactive decay, would be no more than 4,000 years old.

What's interesting is that earth's history of cataclysmic events is not questioned by anyone -- neither scientist nor Biblical scholar. They may differ in their accounts of what occurred, but not necessarily in the severity of the events.

The Bible's account of The Flood, of course, would have been the mother of all catastrophes. It entailed heat, pressure, and an unimaginable mixture of elements. This would certainly have far exceeded any extreme conditions created by scientists in a lab.

The scientific account of earth's formation and development is no less catastrophic:

Earth formed of the debris flung off the sun's violent formation about 4.5 billions years ago. Being a molten planet in it's initial stages, earth's dense materials of molten nickel and iron flowed to the center, and its lighter materials, such as molten silicon, flowed to the top. Eventually, earth cooled and solidified into a core, mantle and crust.

Earth's original atmosphere consisted of Hydrogen and Helium. This atmosphere subsequently heated to escape-velocity by solar radiation and escaped into space. It took about 2 billion years for oxygen to appear in earth's atmosphere, eventually resulting in an atmosphere consisting of 78% Nitrogen and 20% Oxygen.

Our planet has been pounded by meteorites throughout history. One such impact, in Mexico, an alleged 65 million years ago, was so intense that it resulted in mass extinctions, including the extinction of the dinosaur.

Earth has gone through several ice ages. The last one ended around 10,000 years ago, after lasting roughly 60,000 years. At one point 97% of Canada was covered in ice.

The fact is we're detecting natural variations in the rate of radioactive decay even today, in a relative calm period of global and cosmological history. "Recent reports of periodic fluctuations in nuclear decay data of certain isotopes have led to the suggestion that nuclear decay rates are being influenced by the Sun ... " reported the Cornell University website (arxiv.org/abs/1007.3318) on July 20, 2010.

And they're not alone.

* The Atlantic: TheAtlantic.com

(August 25, 2010) "Radioactive elements on Earth are like geological watches. A radioactive isotope of carbon is used to date human civilizations, among other things, because we know that its half-life is precisely 5,730 years; count how much of the carbon 14 has decayed and you can get a pretty accurate measure of how old something is. (If half of the expected amount is left, you'd say, 'This thing is likely 5,730 years old.')

"But what if the rate of radioactive decay -- the watch -- was not constant? One minute, the second hand is moving at one speed, and the next it has sped up or slowed down. And what if what changed that rate of decay was solar activity on the sun, 93 million miles away?

"That's what recent research at Purdue University suggests. In a slate of recent papers, physicists Ephraim Fischbach and Jere Jenkins argue that measured differences in the decay rates of radioactive isotopes cannot be explained by experimental errors. Instead, they seem to vary with the earth's distance from the sun and periodic changes in solar activity."

Ephraim Fischbach is a professor of physics, with a B.A. in Physics from Columbia University and a Ph.D. and M.S. in Physics from the University of Pennsylvania. Jere Jenkins is Director of the Radiation Laboratories at the School of Nuclear Engineering.

   * AstroEngine - AstroEngine.com

      (September 26, 2008) The paper entitled 'Evidence for Correlations Between Nuclear Decay Rates and Earth-Sun Distance' by Jenkins et al. studied the link between nuclear decay rates of several independent silicon and radium isotopes. Decay data was accumulated over many years and a strange pattern emerged; radioactive decay rates fluctuated with the annual variation of Earth's distance from the Sun (throughout Earth's 365 day orbit, our planet fluctuates approximately 0.98 AU to 1.02 AU from the Sun)." [1 AU (Astronomical Unit) is approximately 93 million miles, the distance from earth to the sun.]

Further studies of radioactive material on board spacecrafts, as they moved away from the sun, showed that distance from the sun is not the culprit, and the cause of radioactive variations remains a mystery.

* Stanford University - news.stanford.edu

"It's a mystery that presented itself unexpectedly: The radioactive decay of some elements sitting quietly in laboratories on Earth seemed to be influenced by activities inside the sun, 93 million miles away.

"Is this possible?

"Researchers from Stanford and Purdue University believe it is. But their explanation of how it happens opens the door to yet another mystery.

"There is even an outside chance that this unexpected effect is brought about by a previously unknown particle emitted by the sun. 'That would be truly remarkable,' said Peter Sturrock, Stanford professor emeritus of applied physics and an expert on the inner workings of the sun. 'It's an effect that no one yet understands. Theorists are starting to say, "What's going on?" But that's what the evidence points to. It's a challenge for the physicists and a challenge for the solar people too.'"

Consequently, with a varying radioactive decay rate, there's no way to tell what the radioactive saturation level of any substance or organism was years ago and how long it took for that radioactivity to decay, rendering current dating methods inaccurate and unreliable.